ENTANGLEMENTS

A Wesleyan Chapbook

ENTANGLEMENTS
rae armantrout

Wesleyan University Press
Middletown, Connecticut

Wesleyan University Press
Middletown Connecticut 06459
www.wesleyan.edu/wespress

© 2017 Rae Armantrout
All rights reserved

ISBN 978-0-0815-7739-9

Contents

Note to the Reader
Accounts
Entanglement
Dress Up
Close
Running
Up to Speed
Djinn
Spin
Integer
Inscription
Chirality
Material
Conclusion
Fundamentals
The Ether
Outburst
Making
The Emotional Life of Plants
Hence
So

For John Granger

Note to the Reader

This little book is the result of my years-long interest in reading about science, particularly physics. The poems grew in part from reading books by Brian Cox, Brian Greene, Carlo Rovelli, Nick Lane (who has written on the ways life makes use of "quantum weirdness"), Richard Feynman, Frank Wilczek and many others who have attempted to clarify the discoveries of modern physics for lay people. They were also enriched by conversations with two physicists I've known: Brian Keating and Juan Jose Gomez-Cardena. I called it Entanglements not only for the baffling way two particles can become entangled so that they appear to communicate instantaneously, but also because of the way my daily life experiences and emotions became entangled (in these poems) with what I was learning about physics. I want to dedicate the book to my friend John Granger who first recommended many of these books to me.

ACCOUNTS
for Brian Keating

Light was on its way
from nothing
to nowhere.

Light was all business

    Light was full speed

when it got interrupted.

Interrupted by what?

When it got tangled up
and broke
into opposite

    broke into brand new things.

    What kinds of things?

    Drinking Cup

    "Thinking of you!
                Convenience Valet"

How could speed take shape?

*

Hush!
Do you want me to start over?

*

The fading laser pulse

        Information describing the fading
                  laser pulse

is stored

    is encoded

in the spin states
of atoms.

God
is balancing his checkbook

    God is encrypting his account.

This is taking forever!

ENTANGLEMENT

1

"Don't let the car fool you.

My treasure
is in heaven."

2

The material world is made up
entirely

of collisions

between otherwise
indefinite objects.

Then what is a collision?

(Or the physical world

collapses

into place

at the shock of
being seen.)

3

In the shorter version,

tentacled
stomach swallows stomach.

In a long dream,
I'm with Aaron,

visiting his future,
helping him make choices.

DRESS UP

To be "dressed"
is to emit
"virtual particles."

*

The spirit of "renormalization" is that

an electron
all by itself

can have infinite
mass and charge,

but, when it's "dressed"…

*

A toddler stares at us
 till we look up.

"Flirtatious," we call it.

She waits
until we get the joke

about being here,
being there.

CLOSE

1

As if a single scream
gave birth

to whole families
of traits

such as "flavor," "color,"
"spin"

and this tendency to cling.

2

Dry, white frazzle
in a blue vase –

beautiful –

a frozen swarm
of incommensurate wishes.

3

Slow, blue, stiff
are forms

of crowd behavior,

mass hysteria.

Come close.

The crowd is made of
little gods

and there is still
no heaven

RUNNING

Let's say the universe
is made of strings
that "vibrate" or thrash
 in an effort

to minimize the area
that is the product
of their length
and their duration in time.

*

Let's call contraction
"focus"
or "pleasure."

*

You'll step forward,
I know,

into the contracting
light

ready to like
anyone.

How far will you get?

You'll be far ahead
and distracted.

By what?

I won't see it.

I'll be running to catch up.

I'll know you
by your willingness.

I won't believe

that what's continual
is automatic

UP TO SPEED

Streamline to instantaneous
voucher in/voucher out
system.

The plot winnows.

The Sphinx
wants me to guess.

Does a road
run its whole length
at once?

Does a creature
curve to meet
itself?

Whirlette!

*

Covered or cupboard
 breast? Real

housekeeping's
kinesthesiac. Cans

held high
to counterbalance "won't."

Is it
such agendas

which survive
as souls?

*

Vagueness is personal!

A wall of concrete bricks,
right here,
while sun surveys its grooves

and I try
"instantly" then "forever."

But the word is
way back,
show-boating.

Light is "with God"

(light, the traveler).

*

Are you the come-on
and the egress?

One who hobbles by
determinedly?

Not yet?

DJINN

Haunted, they say, believing
the soft, shifty
dunes are made up
 of false promises.

Many believe
whatever happens
 is the other half
of a conversation.

Many whisper
white lies
to the dead.

"The boys are doing really well."

Some think
nothing is so
until it has been witnessed.

They believe
the bits are iffy;

the forces that bind them,
absolute.

SPIN

That we are composed
of dimensionless points

which nonetheless spin,

which nonetheless exist
in space,

which is a mapping
of dimensions.

*

The pundit says
the candidate's speech
hit
"all the right points,"

hit "fed-up" but "not bitter,"
hit "not hearkening back."

*

Light strikes our eyes
and we say, "Look *there*!"

INTEGER

1

One what?

One grasp?

No hands.

No collection

of stars. Something dark

pervades it.

2

Metaphor
is ritual sacrifice.

It kills the look-alike.

No,
metaphor is homeopathy.

A healthy cell
exhibits contact inhibition.

3

These temporary credits
will no longer be reflected
in your next billing period.

4

"Dark" meaning

not reflecting,

not amenable
to suggestion.

INSCRIPTION

God
as the lace-making
machine,

the hypnotized spider.

Why shouldn't
an idée fixe
be infinite?

Blithering symmetries.

More of you are coming.

*

"I think our incentives
are sexy and edgy."

*

As if you
could become another person
by setting off
an automatic

cascade of responses
in his/her body.

As if you could escape
by following

the path you carved
there

to its prescribed end.

*

Poems addressed
to their own dead letters –
campy femme-fatales.

Poems addressed
to their end-times'

desiccation.

Entropy increases as I recall
less and less
of the number string.

Snackle-crackle
of strings breaking –
that radiation hiss evening things out.

*

Look – I'm cooperating!
I can pull myself apart
and still speak

CHIRALITY

If I didn't need
to do anything,
would I?

Would I oscillate
in two
or three dimensions?

Would I   summon
a beholder

and change chirality
for "him?"

A massless particle
passes through the void
with no resistance.

Ask what it means
to pass through the void.

Ask how it differs
from not passing.

MATERIAL

Oh, you're wearing the gold one.

That's my favorite,
to be honest.

The gun metal is all gone.

*

Packet.
Pocket.
Point.

For us to consist
of infinitesimal points

of want
and not

makes a lot of sense.

(For a point to consist
of the array

of its own
possible locations

For locations to consist

For "consist" to consist
of a pair
of empty pockets)

*

Here, you try it on.

CONCLUSION

1

A man is upset for many years
because he's heard
that information is destroyed
in a black hole.

Question: what does this man mean
by "information?"

The example given
 is of a cry for help,

but this is accompanied
by the image of a toy space ship,
upended,

and is thus
not to be taken seriously.

The man recovers his peace of mind
when he ceases to believe
in passing through,

when he becomes convinced
that the lost information

is splattered
on the event

horizon.

 2

The detective is the new mime.

She acts out understanding
the way a mime
climbs an invisible wall.

*

It's because our senses
are so poor that,

on CSI,
the investigators
stand stock---still,

boulders in a stream,

while a crowd
pours around them.

They pan
in slow motion, reminding us
of cameras,

then focus
with inhuman clarity

on the pattern of cracks
in a wall.

3

God's fractal
stammer

pleasures us
again.

FUNDAMENTALS

Why is it that
for it

to be in-
finitely large

is terrific,
but to be

infinitely small
is just

unthinkable?

The thought
of a smaller

bit inside
each bit

goes nowhere
still

has symmetry
going on

and on
about it.

Then there's our model
in which

the fundamentals
are sound,

impenetrable nubs

THE ETHER

We're out
past the end

game where things
get fuzzy,

less thingy,

though in past times
we practiced

precision
concrete as a slot machine.

But to be precise
you need to stop

a moment
which turns out to be

impracticable
and besides

speed is of the essence.

Don't worry.

"Of" can take care
of itself

and it's fine
to say "essence"

now that it's understood
to mean ether,

a kind of filler
made either

of inattention
or absorption

somewhere near
the Planck length

OUTBURST

1

What do you like best
about the present?

Reflection---

its spangle
and its non-
locality:

those eucalyptus leaves
as points of light

splashed
over this windshield.

2

What if every moment
is a best guess
on a pop quiz?

As if waking up,
I stop explaining

Tony Soprano's outburst

to his aggravated
henchmen.

MAKING

"What made this happen?"
you ask every time

as if
compulsion itself
were mandatory,

the way light travels
at the speed of light

"because it must"

*

It is in no sense
essential

that this crown of leaves,
sifted by wind

as if turning over
some problem,

is a grey-green

brightening into rust-red

at the tips

or that its equivocations
fill this instant

to the brim.

*

Whille light
 has caught up

to itself
again

and only seems
to be making

time

THE EMOTIONAL LIFE OF PLANTS

An exciton consists
of the escaped negative
(electron)
and the positive hole
it left behind.

This binary system
is unstable
and must be transported quickly
to the processing center
which, in practice, means
the exciton must be left
strictly alone
so that it travels
all ways at once
going nowhere
but also
going directly
to the factory floor.

Leaving aside the question
of what it means
for a positive hole
to be "left behind"
and also to travel
as half of a system,
this happens because nothing
can be still
and because, for the lonely,
direction is meaningless.

HENCE

Sculpted minarets
of clouds gone

hence – or thence?

No dreams
are that well formed.

*

No one can depict

the absolute bracelets
of the orbits

touched upon
by electrons

as feeling dithers
between words

SO

So that nothing
rhymes with much --

or starts to
and thinks better of it.

*

Ending with "like"
or "so."

Ending with "as if"

*

Virtual particles
carry the current

Rae Armantrout is the author of thirteen books of poetry including *Partly: New and Selected Poems, 2001-2015* and *Versed*, winner of the Pulitzer Prize for poetry and the National Book Critics Circle Award.

Poems in this chapbook first appeared in the following Wesleyan University Press books:
Up to Speed (2004): Entanglement, Up to Speed
Next Life (2007): Close
Versed (2009): Running, Integer, Inscription
Just Saying (2013): Accounts, Dress Up
Money Shot (2011): Djinn, Spin
Itself (2015): Chirality, Material, Fundamentals
Partly: New and Selected Poems, 2001-2015 (2016): The Ether, Outburst
Wobble (2018): Making, The Emotional Life of Plants, Hence, So

Wesleyan Chapbooks

Entanglements, Rae Armantrout

I Will Teach You about Murder: 29 Love Poems, edited by Shea Fitzpatrick, Sallie Fullerton and Torii Johnson

CPSIA information can be obtained
at www.ICGtesting.com
Printed in the USA
BVOW11s1127080617
485856BV00005BA/10/P